HF269691

Solstices

by Grace Hansen

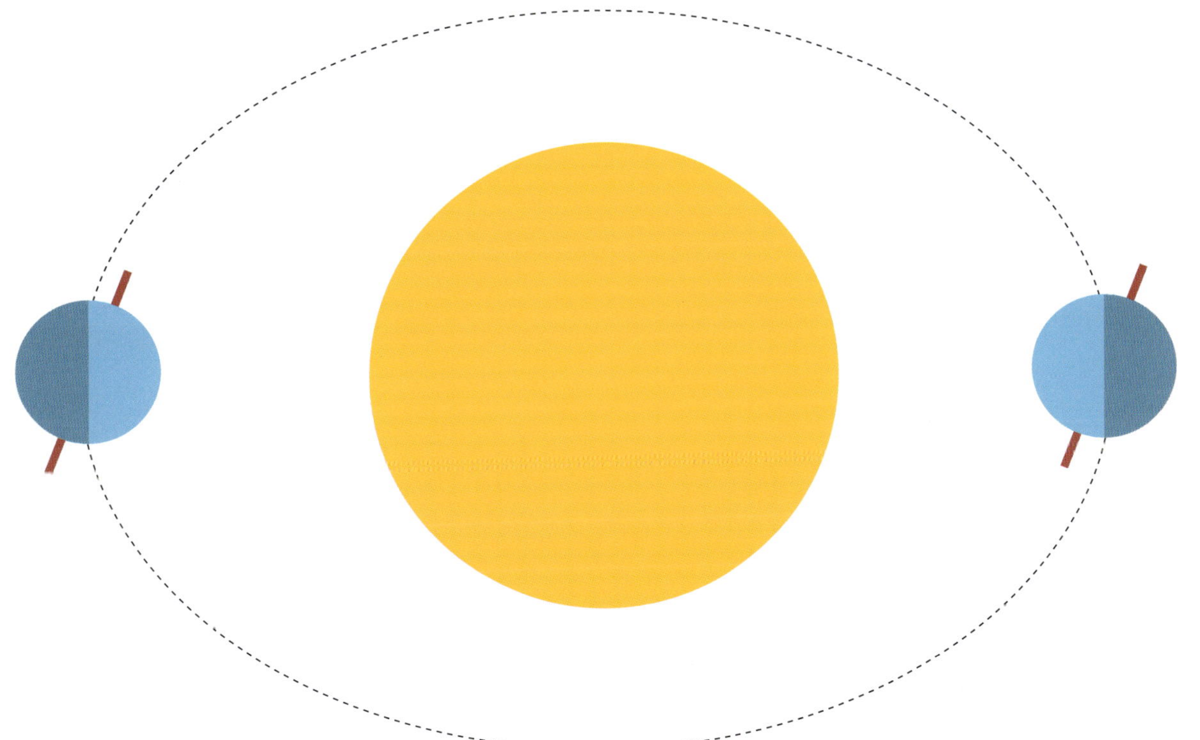

abdobooks.com

Published by Abdo Kids, a division of ABDO, P.O. Box 398166, Minneapolis, Minnesota 55439.
Copyright © 2020 by Abdo Consulting Group, Inc. International copyrights reserved in all countries.
No part of this book may be reproduced in any form without written permission from the publisher.
Abdo Kids Jumbo™ is a trademark and logo of Abdo Kids.

Printed in the United States of America, North Mankato, Minnesota.

102019

012020

Photo Credits: iStock, NASA, Shutterstock, Shutterstock PREMIER p.21

Production Contributors: Teddy Borth, Jennie Forsberg, Grace Hansen
Design Contributors: Dorothy Toth, Pakou Moua

Library of Congress Control Number: 2019941254
Publisher's Cataloging-in-Publication Data

Names: Hansen, Grace, author.

Title: Solstices / by Grace Hansen

Description: Minneapolis, Minnesota : Abdo Kids, 2020 | Series: Sky lights | Includes online resources
    and index.

Identifiers: ISBN 9781532189111 (lib. bdg.) | ISBN 9781532189609 (ebook) | ISBN 9781098200589
    (Read-to-Me ebook)

Subjects: LCSH: Solstice, Summer--Juvenile literature. | Solstice, Winter--Juvenile literature. | Earth
    (Planet)--Rotation--Juvenile literature. | Seasons--Juvenile literature. | Sun--Juvenile literature.

Classification: DDC 525.5--dc23

Table of Contents

Solstices

Earth experiences two solstices each year. They are the summer and winter solstices.

summer
solstice
June 20-21

winter
solstice
December 21-22

The summer solstice is the day of the year with the most daylight. The winter solstice is the day of the year with the least daylight.

Chicago
5:00 pm
June
Chicago
5:00 pm
December

Solstices happen because Earth

orbits the sun. Earth is also

tilted on its **axis**.

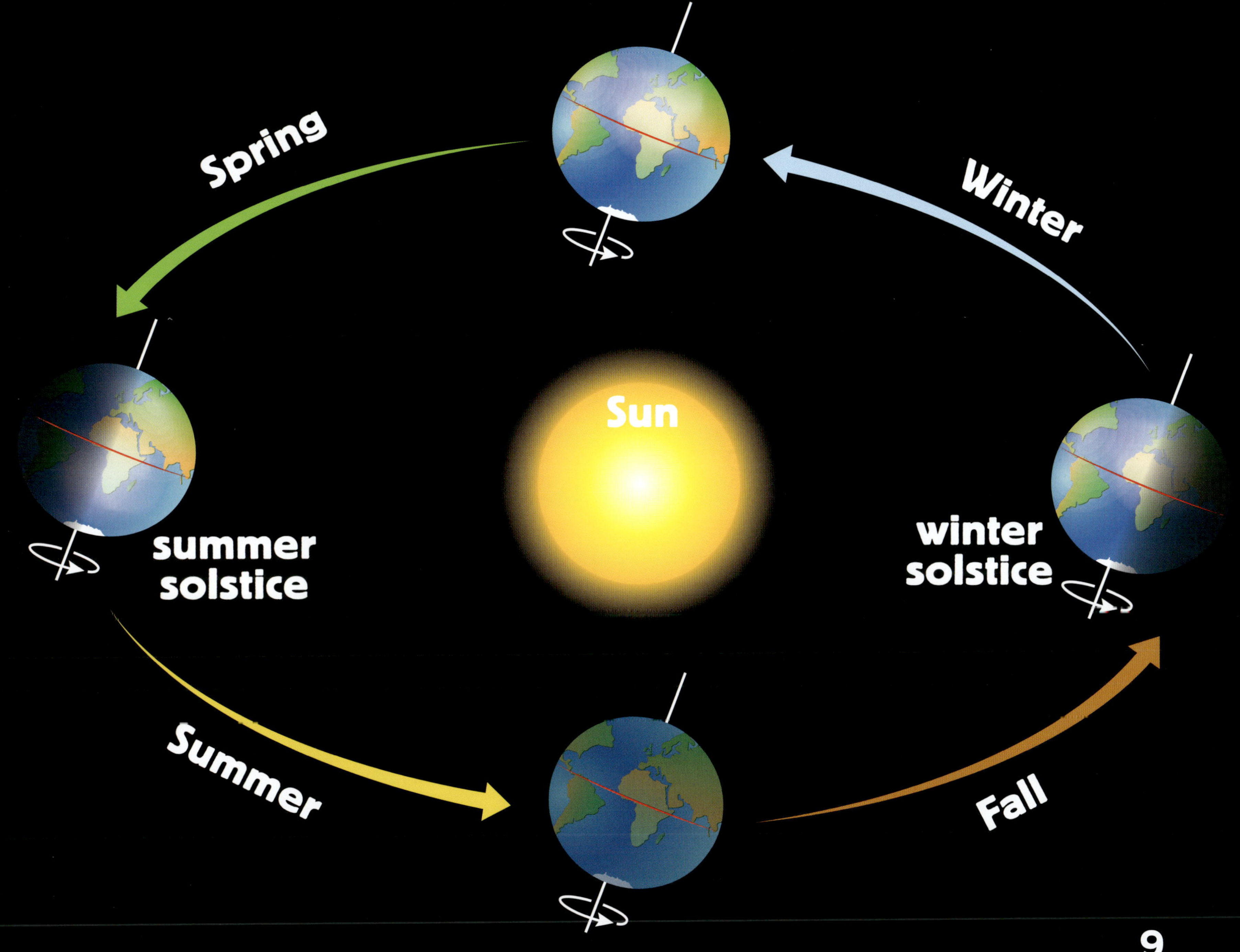

Spring
Winter
Sun
summer
solstice
winter
solstice
Summer
Fall

Earth is divided into two

hemispheres.

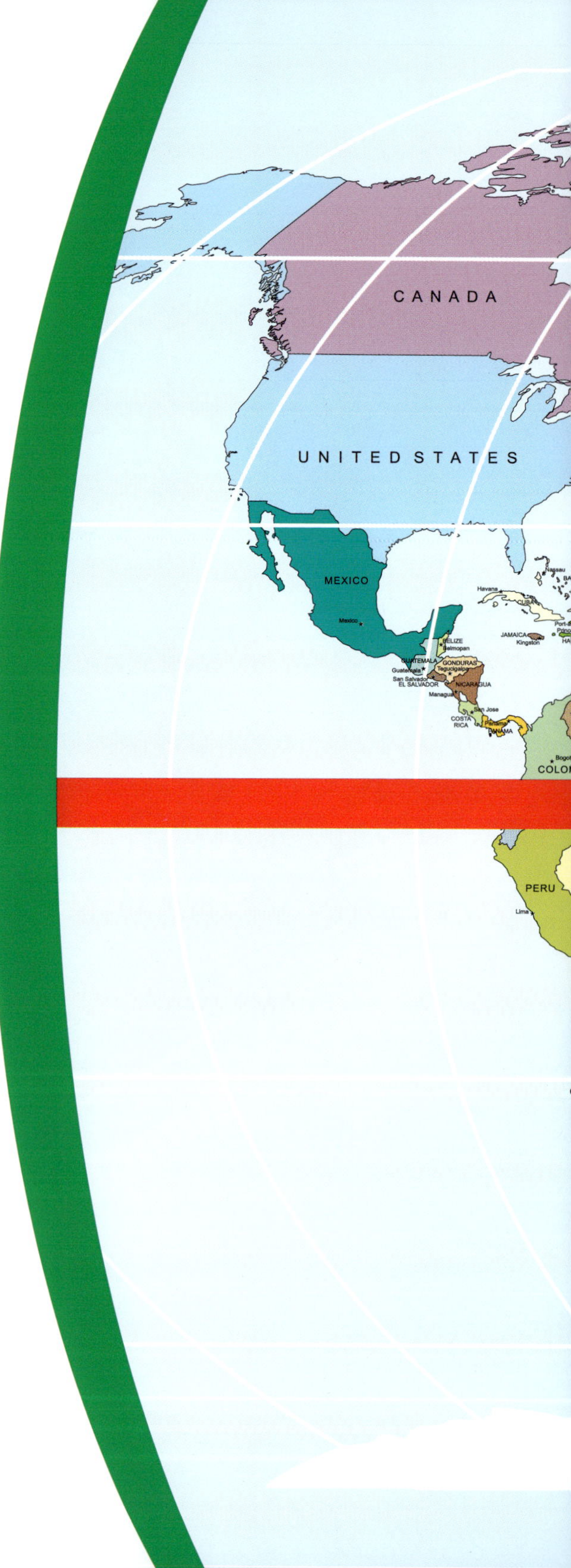

GREENLAND
Nuuk
ICELAND
Reykjavik
NORWAY
SWEDEN
FINLAND
Stockholm
Helsinki
EST
LAT
LITH
Minsk
Moscow
RUSSIA
UNITED KINGDOM
Dublin
IRELAND
NETH
GERMANY
Berlin
Warsaw
POLAND
BELARUS
Kyiv
UKRAINE
KAZAHSTAN
Astana
Ulaanbaatar
MONGOLIA
FRANCE
Paris
SWITZ
Ljubljana
HUNGARY
Budapest
Chisinau
ROMANIA
MOLDOVA
Bucharest
GEORGIA
ARMENIA
AZERBAIJAN
Baku
Bishkek
KYRGYZSTAN
Beijing
NORTH KOREA
Pyongyang
JAPAN
Tokyo
PORTUGAL
Lisbon
SPAIN
Madrid
ITALY
MALTA
Valletta
TUNISIA
Tunis
GREECE
Ankara
TURKEY
SYRIA
Yerevan
Tbilisi
UZBEKISTAN
TURKMENISTAN
Ashgabat
Dushanbe
TAJIKISTAN
Tehran
Kabul
SOUTH KOREA
CHINA
NORTHERN HEMISPHERE
MAURITANIA
Nouakchott
MALI
NIGER
Niamey
CHAD
N'Djamena
ARABIA
Muscat
UNITED ARAB EMIRATES
OMAN
INDIA
BURMA
LAOS
Hanoi
Vientiane
CAPE VERDE
Praia
SENEGAL
Dakar
THE GAMBIA
Banjul
GUINEA-BISSAU
Bissau
GUINEA
Conakry
Bamako
BURKINA FASO
Ouagadougou
BENIN
NIGERIA
Abuja
SUDAN
Khartoum
ERITREA
Asmara
YEMEN
THAILAND
Bangkok
VIETNAM
Manila
PHILIPPINES
SIERRA LEONE
Freetown
LIBERIA
Monrovia
COTE D'IVOIRE
Yamoussoukro
GHANA
Accra
Porto-Novo
CENTRAL AFRICAN REPUBLIC
Bangui
ETHIOPIA
Addis Ababa
DJIBOUTI
Djibouti
CAMBODIA
Phnom Penh
CAMEROON
Yaounde
Malabo
SOMALIA
MALDIVES
Male
SRI LANKA
Colombo
Bandar Seri Begawan
BRUNEI
EQUATOR
REPUBLIC OF THE CONGO
Brazzaville
Kinshasa
BURUNDI
Bujumbura
TANZANIA
Dar es Salaam
INDONESIA
Jakarta
PAPUA NEW GUINEA
Port Moresby
SOLOMON ISLANDES
Honiara
BRAZIL
Luanda
ANGOLA
ZAMBIA
Lusaka
Lilongwe
MOZAMBIQUE
ZIMBABWE
Antananarivo
MADAGASCAR
Port Louis
MAURITIUS
VANU
Port-Vil
NAMIBIA
Windhoek
BOTSWANA
Gaborone
NEW CALEDONIA
Noumea
GUYANA
Georgetown
SURINAME
Paramaribo
French Guiana (FRANCE)
Cayenne
AUSTRALIA
Canberra
SOUTHERN HEMISPHERE
VIA
Brasilia
PARAGUAY
Asuncion
URUGUAY
Buenos Aires
Montevideo
ARGENTINA
NEW ZEALAND
FALKLAND ISLANDS
Stanley
ANTARCTICA

Earth has two poles. The North Pole is the northernmost part of Earth. The South Pole is the southernmost part of Earth.

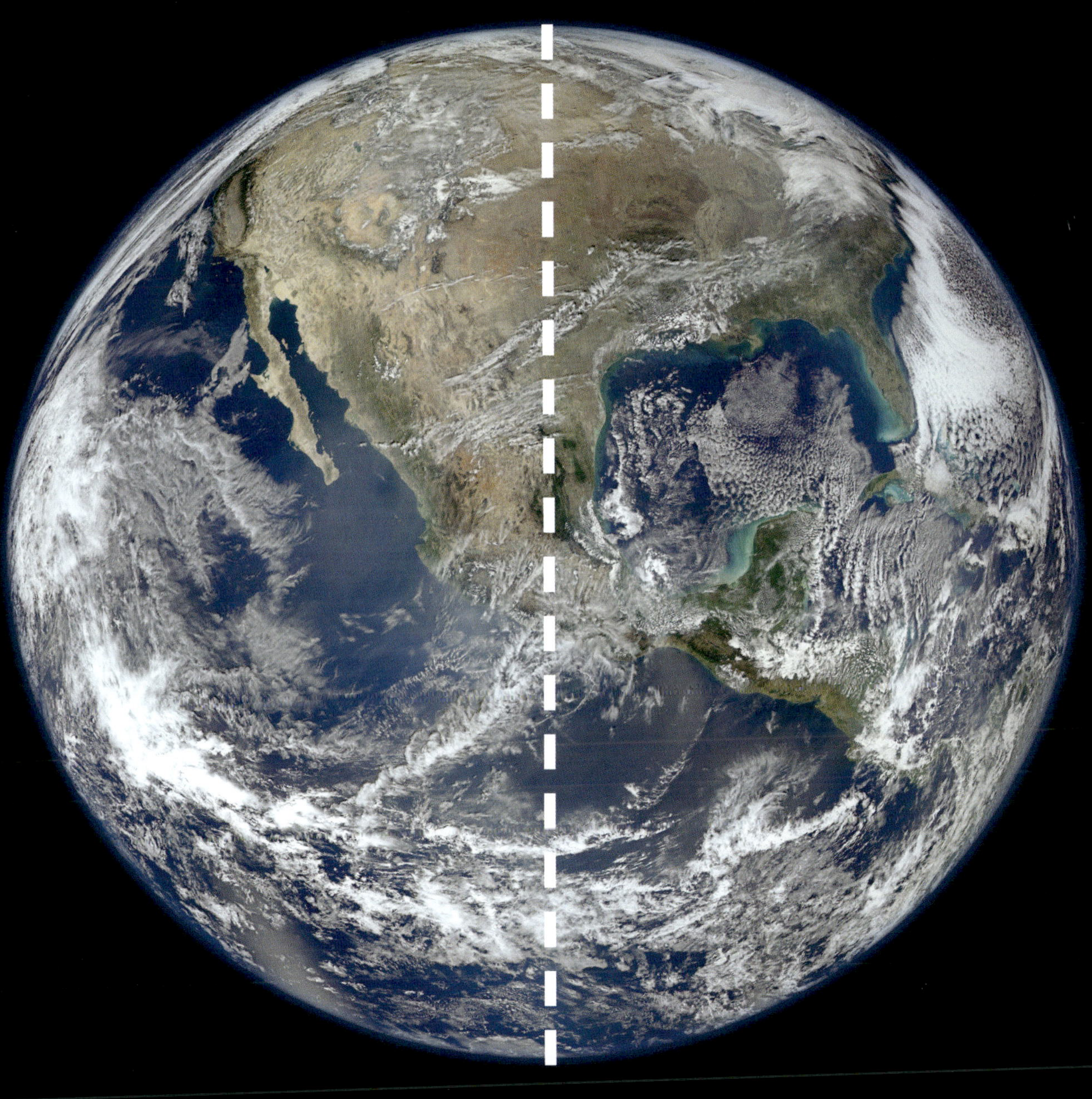

North Pole
South Pole

On a day in June, the North Pole is most directly pointed toward the sun. This marks the summer solstice in the northern **hemisphere**. It is the day with the longest and brightest daylight.

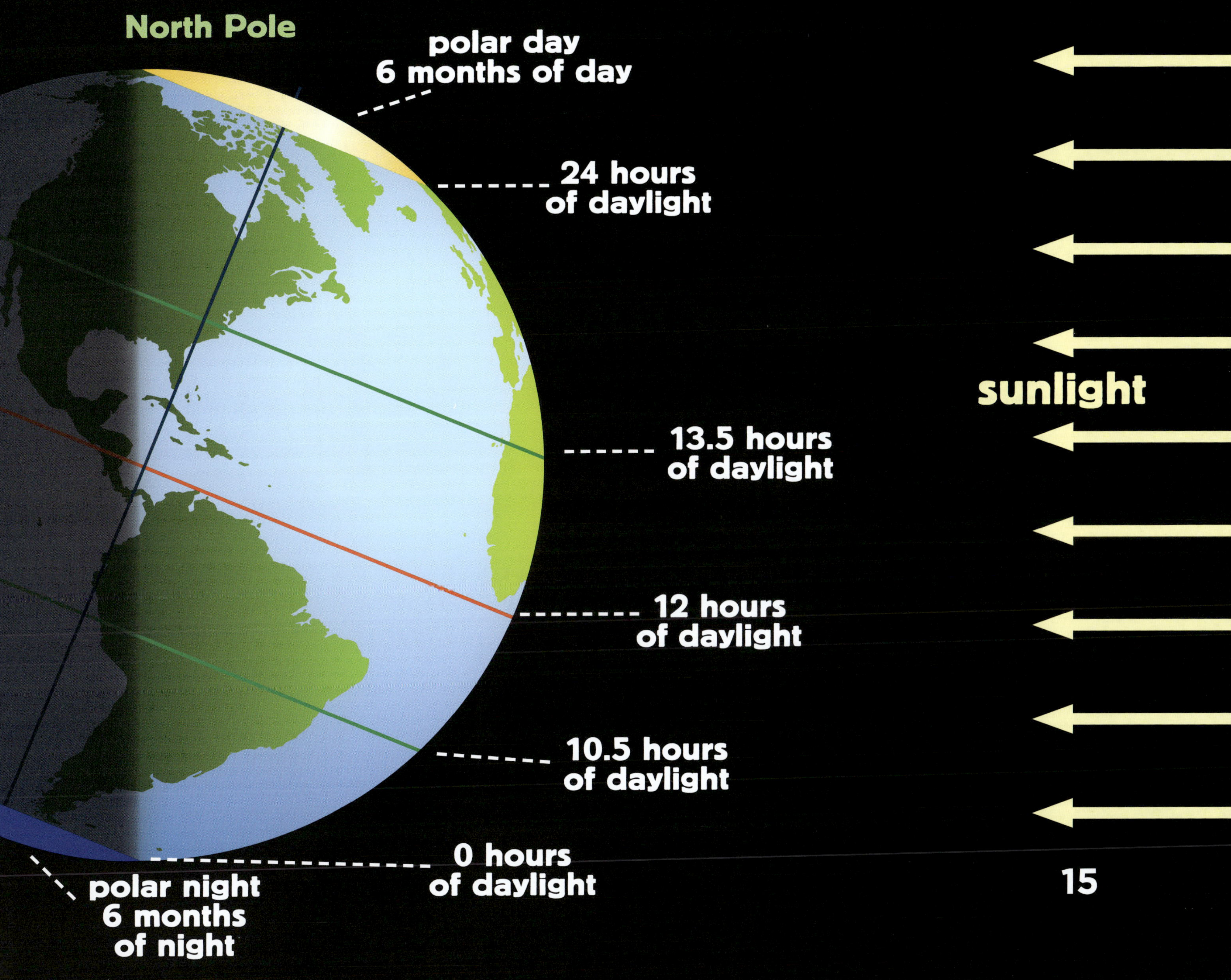

North Pole
polar day
6 months of day
24 hours
of daylight
13.5 hours
of daylight
12 hours
of daylight
10.5 hours
of daylight
0 hours
of daylight
polar night
6 months
of night
sunlight
15

On a day in December, the
North Pole is most directly
pointed away from the sun.
This marks the winter solstice
in the northern **hemisphere**. It
is the day with the shortest and
darkest daylight.

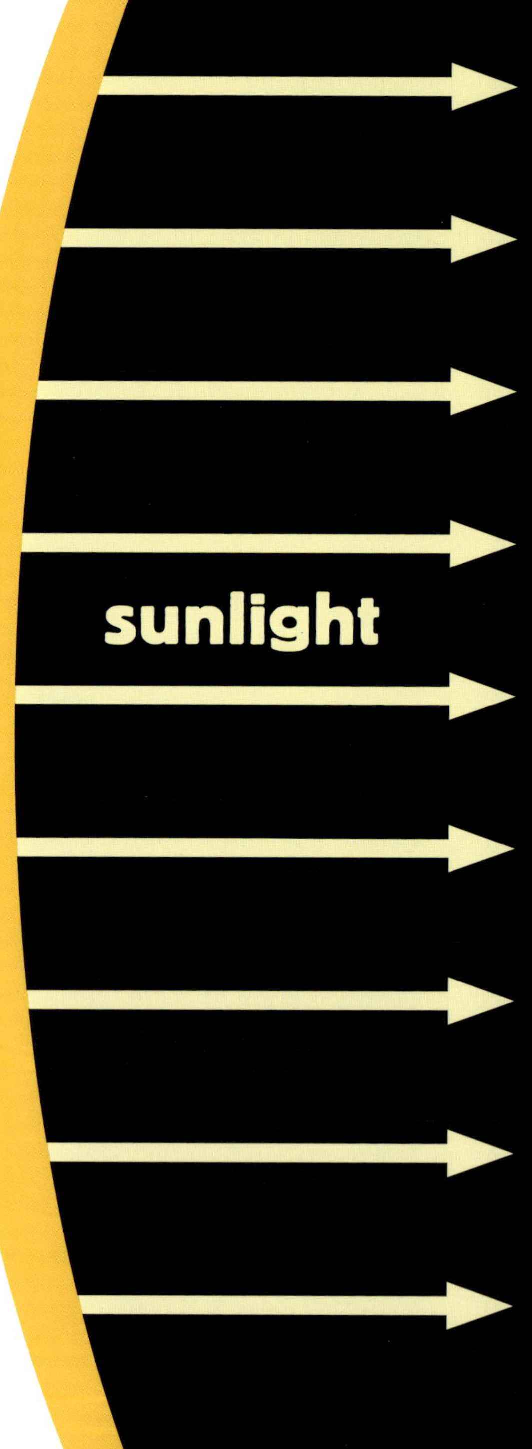

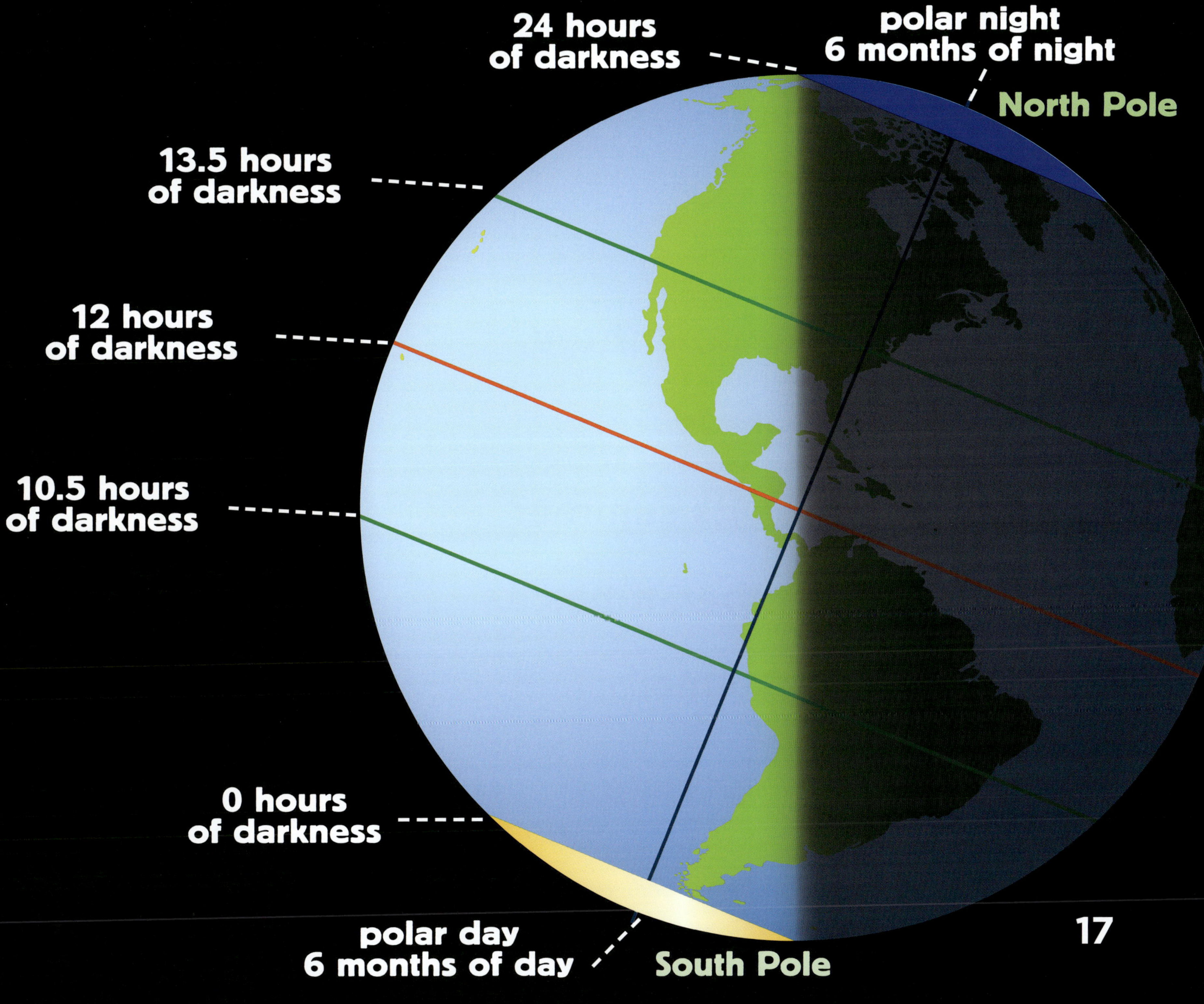

24 hours of darkness
13.5 hours of darkness
12 hours of darkness
10.5 hours of darkness
0 hours of darkness
polar night 6 months of night
North Pole
polar day 6 months of day
South Pole
17

In the southern **hemisphere**, the exact opposite is happening. It has its summer solstice in December. It has its winter solstice in June.

19

Solstice Celebration

Every year, many people gather at **Stonehenge** in England. When the sun rises on the summer solstice, sunlight is channeled into the **monument's** center.

More Facts

- The word solstice comes from the Latin language. *Sol* means "sun" and *sistere* means "to stand still."

- In Alaska, a third of the state sits north of the Arctic Circle. This is an area where the sun does not set on the summer solstice. Alaskans celebrate with festivals and even a midnight baseball game!

- **Stonehenge** is also aligned to the sunset on the winter solstice.

Glossary

axis – an imaginary line through the center of Earth, around which it turns.

hemisphere – either of two halves of the Earth. A hemisphere is formed by dividing Earth into the Northern and Southern hemispheres at the equator.

monument – a structure built to celebrate a person or event.

orbit – the curved path of a planet, moon, or other object around a larger celestial body.

Stonehenge – a circular group of stones in Southern England that was built around 5,000 years ago.

Index

Visit **abdokids.com** to access crafts, games, videos, and more!